Any mistakes in the information, descriptions, etcetera, are mine entirely. Despite extensive research, there can sometimes be contradictory information available, which makes it challenging to filter the accuracy or inaccuracy of such information. Although the publisher and the author have made every effort to ensure that the information in this book was correct at the time of press, and while this publication is designed to provide accurate information regarding the subject matter covered, the publisher and the author assume no responsibility for errors, inaccuracies, omissions, or any other inconsistencies herein.

ISBN: 978-1-7641194-8-1 (Paperback)

First edition November 2025
A CONSERVATION CLUB PUBLICATION
www.conservationclub.com.au

NATIONAL LIBRARY OF AUSTRALIA

A catalogue record for this book is available from the National Library of Australia

In the spirit of reconciliation, Conservation Club acknowledges the Traditional Custodians of Country throughout Australia and their connection to land, sea, and community. We pay our respects to their Elders, past and present, and extend that respect to all Aboriginal and Torres Strait Islander peoples.

To my grandchildren, Samuel, Jameson, Harry and Audrey.

I hope that you all enjoy the same sense of wonder and love of the environment that I have as you grow into bigger children, young adults and into adulthood and that you have the opportunity to explore many different environments during your lifetimes. Perhaps you will become warriors for wildlife in the future.

In the forest, along shorelines, in wetlands, by the creek,
Birds are squawking and chattering.... chirp, trill, peep, cheep!
A **King Parrot** calls out, “Come with me, take a peek -
You can tell who eats what from the shape of the beak!”

It’s all about the beak, about the bill, about the beak,
Which tells you whether we poke, prod, tear, twist or tweak.
Some of us peck, pluck, sift, tear and sip,
Some of us sweep, scoop, crush, pry, flip or nip!

Can you tell from the shape of the king parrot’s beak what it feeds on? I think it has the type of beak that crushes seeds and seed pods... What do you think?

The **Australian Pelican** grins wide as wide can be,
"My beak's like a bucket for fish from the shining sea.
I scoop up silvery snacks when I plunge or glide and seek...
A bucket built for ocean feasts, the perfect pelican beak!"

It's all about the beak, about the bill, about the beak,
Which tells you whether we poke, prod, tear, twist or tweak.
Some of us peck, pluck, sift, tear and sip,
Some of us sweep, scoop, crush, pry, flip or nip!

It is pretty obvious how the pelican feeds and what it feeds on, don't you think?

BUT.....did you know that the Australian pelican holds the world record for the longest beak AND, it also uses it's beak pouch to cool down by opening the beak wide and rapidly vibrating the pouch with each breath!?

The **Powerful Owl** blinks twice, then gives a booming hoot,
"My hooked beak holds dinner when my silent wings commute.
I glide through shadowed forests where the sleepy possums sneak,
I'm built for hunting in the dark, from talons to my beak!"

It's all about the beak, about the bill, about the beak,
Which tells you whether we poke, prod, tear, twist or tweak.
Some of us peck, pluck, sift, tear and sip,
Some of us sweep, scoop, crush, pry, flip or nip!

How do you think the powerful owl feeds and hunts? I think that his beak is built for tearing flesh... Do you agree?

..................I am glad I am not a possum!

The **Ruddy Turnstone** walks the shore, flipping stones and sand,
Its stout beak lifts the seaweed mats the tides throw on the land.
Hidden crabs and wriggling worms appear when rocks are turned -
A clever bird that finds its food in lessons it has learned.

It's all about the beak, about the bill, about the beak,
Which tells you whether we poke, prod, tear, twist or tweak.
Some of us peck, pluck, sift, tear and sip,
Some of us sweep, scoop, crush, pry, flip or nip!

How very smart of the turnstone – now I understand why this bird is called a turnstone – do you?

The **Red-browed Finches** chirp, “Our beaks are short and strong,
They crunch through seeds and grains we’ve searched for all day long.
We love the nuts and tiny seeds that fall where branches creak -
Our sturdy, conical crunching tools are perfect finch-style beaks!”

It’s all about the beak, about the bill, about the beak,
Which tells you whether we poke, prod, tear, twist or tweak.
Some of us peck, pluck, sift, tear and sip,
Some of us sweep, scoop, crush, pry, flip or nip!

The red-browed finch often feeds in small flocks of 10- 20 birds.
They are very sociable – PARTY ON!

The **Wedge-tailed Shearwater** skims the waves with slender bill,
It's made for contact-dipping hunts that need both speed and skill.
Its tube-tipped nose drains salty drops from every ocean sip—
A tubenose sailor of the sea on long, far-flying trips.

It's all about the beak, about the bill, about the beak,
Which tells you whether we poke, prod, tear, twist or tweak.
Some of us peck, pluck, sift, tear and sip,
Some of us sweep, scoop, crush, pry, flip or nip!

Wedge-tailed shearwaters usually hunt by snatching prey from the water's surface while flying low. Sometimes they will form large flocks to feed on fish driven to the surface by other predators and can do this when skimming or when sitting floating on the surface. They can also plunge into the water, diving down usually to between 5 and 14 metres, although they have been recorded to dive much deeper!

The **Sulphur-crested Cockatoo** laughs loud, "I pry and twist,
My strong curved beak cracks seeds that most beaks can't resist.
I use it like a nutty tool for any snack I seek—
A mighty crunching, cracking clamp, my cockatoo-style beak!"

It's all about the beak, about the bill, about the beak,
Which tells you whether we poke, prod, tear, twist or tweak.
Some of us peck, pluck, sift, tear and sip,
Some of us sweep, scoop, crush, pry, flip or nip!

Sulphur-crested cockatoos forage in groups, mainly in the mornings and evenings, for a natural diet of seeds, nuts, berries, roots, and insects. They use their powerful beaks to shell hard nuts and also chew on branches and leaves to keep their beaks trimmed! They can also be a nuisance to farmers, feeding on crops such as mangos, strawberries, rockmelon, and grapes. A bit cheeky!

The **Red-necked Avocet** sweeps the shallows at the tide,
Its upturned beak stirs tiny prey that try so hard to hide.
It swishes side to side where little creatures creep—
A dancer on the waterline, its rhythm smooth and deep.

It's all about the beak, about the bill, about the beak,
Which tells you whether we poke, prod, tear, twist or tweak.
Some of us peck, pluck, sift, tear and sip,
Some of us sweep, scoop, crush, pry, flip or nip!

Avocets feed on small crustaceans like brine shrimp and seed shrimp, small molluscs like aquatic snails and bivalves, and aquatic insects as well as supplementing their diet with seeds.

The **Black Honeyeater** hums, "My beak is slender, long,
For sipping nectar deep where blossoms drip their song.
I dip into each flower bell for every honeyed sip—
A perfect nectar-tasting straw at every sugary drip!"

It's all about the beak, about the bill, about the beak,
Which tells you whether we poke, prod, tear, twist or tweak.
Some of us peck, pluck, sift, tear and sip,
Some of us sweep, scoop, crush, pry, flip or nip!

Black Honeyeaters feed on nectar from tubular-shaped flowers, especially those of the emu-bush or clumps of mistletoe found high in the crowns of eucalypt trees, but sometimes in jacaranda and poinciana trees.

DID YOU KNOW - Australia has around 90 native mistletoe species? That is a lot of choices on the honeyeater menu!!

Along the shore, **Sharp-tailed Sandpipers** stride through shining mud,
Their beaks probe soft, wet places after each receding flood.
Like tweezers searching quietly, they pinch each worm they seek—
Long, slender, probing hunting tools: the perfect searching beak.

It's all about the beak, about the bill, about the beak,
Which tells you whether we poke, prod, tear, twist or tweak.
Some of us peck, pluck, sift, tear and sip,
Some of us sweep, scoop, crush, pry, flip or nip!

Sandpipers migrate up to 11,000km from their breeding grounds in eastern Russia!

Bristleworms are one of the marine worms that sandpipers like to feed on! Some bristleworms have hollow bristles filled with poison and can cause significant pain and inflammation if they embed in skin, so be sure not to handle these spiky little critters if you find them!

The **Brown Goshawk** swoops low from high in open sky,
"My hooked beak rips meat while other creatures fly.
I'm built for speed and sudden strikes when smaller birds are weak—
A fierce and focused hunter with a sharp and slicing beak!"

It's all about the beak, about the bill, about the beak,
Which tells you whether we poke, prod, tear, twist or tweak.
Some of us peck, pluck, sift, tear and sip,
Some of us sweep, scoop, crush, pry, flip or nip!

Goshawks are opportunistic hunters – they will feed on birds, reptiles and sometimes even dead animals and insects. The European rabbit, an introduced species, is a particularly important prey item!

The **Royal Spoonbill** sways its beak, so wide and flat,
"From side to side, I sweep the shallows where the fish are at!
My spoon-shaped bill snaps up the prey in waters still and deep -
A scooping paddle made for food where muddy currents creep."

It's all about the beak, about the bill, about the beak,
Which tells you whether we poke, prod, tear, twist or tweak.
Some of us peck, pluck, sift, tear and sip,
Some of us sweep, scoop, crush, pry, flip or nip!

The Royal Spoonbill has a black beak and grows some fabulous head feathers when breeding! The Yellow-billed Spoonbill has a similar way of feeding, using its beak to sweep through shallow water in a side-to-side motion to locate prey! When prey is found, the bill snaps shut, and the birds lift their heads to swallow the food.

The **Rainbow Bee**-eater streaks by, flashing colours bright,
Snatching buzzing bees mid-air in perfect, darting flight.
It beats the sting on branch or wire, then eats its snack complete—
A sparkling hunter of the breeze with nimble wings and beak so neat.

It’s all about the beak, about the bill, about the beak,
Which tells you whether we poke, prod, tear, twist or tweak.
Some of us peck, pluck, sift, tear and sip,
Some of us sweep, scoop, crush, pry, flip or nip!

We only have one species of rainbow bee-eater in Australia, although there are around 30 other species that occur from Southern Europe, Africa, through Asia to Papua New Guinea!

The **Eastern Curlew** whispers, "Mine is long and slightly curled,
I poke through mud for worms and snails beneath the watery world.
Along the tidal flats, I search for things that leak and peek—
A careful, poking, probing tool—my curlew-curving beak."

It's all about the beak, about the bill, about the beak,
Which tells you whether we poke, prod, tear, twist or tweak.
Some of us peck, pluck, sift, tear and sip,
Some of us sweep, scoop, crush, pry, flip or nip!

A **Blue-winged Shoveler** duck floats past with gentle quack,
"My wide, flat beak strains tiny snacks and sends the water back.
With comb-like edges on each side I filter what I seek—
A tidy straining, draining tool, the perfect shovel-beak!"

It's all about the beak, about the bill, about the beak,
Which tells you whether we poke, prod, tear, twist or tweak.
Some of us peck, pluck, sift, tear and sip,
Some of us sweep, scoop, crush, pry, flip or nip!

This duck is actually a male mallard, but many ducks have these broad, flat beaks.

Swans, geese and flamingos also have beaks with these comb-like edges or serrations, which are called lamellae.

Now all the birds begin to talk, each proud of beak and bill,
From giant scoops to tiny tips, each shaped to match its skill.
So many forms in one bright world, each suited to its task—
A bird's beak tells its story if you simply stop and ask.

It's all about the beak, about the bill, about the beak,
It shows if we poke, prod, tear, twist or tweak.
Some of us peck, pluck, sift, tear and sip,
Some of us sweep, scoop, crush, pry, flip or nip!

So next time you see a bird that chirps, or trills, or squeaks—
Look closely now and you will see... it's all about the beak!

Teacher & Parent Notes

It's All About the Beak! encourages children to look closely at nature and notice how animals are shaped for the lives they live.
Through playful rhyme and repetition, this story introduces young readers to the **idea of adaptation** — how a bird's beak is perfectly suited to its food and habitat.

As children listen, read, or sing along, they:

- Build **early science knowledge** by connecting beak shape to feeding behaviour.
- Strengthen **vocabulary** through fun, descriptive action words (poke, pluck, pry, nip!).
- Explore **rhyme, rhythm, and repetition** — key foundations for reading fluency.
- Practice **observation and curiosity**, noticing details in the natural world.

This book supports **Early Years and Primary Science** outcomes relating to:

- Animal features and needs
- Habitats and food sources
- How living things adapt to their environment

Encourage children to:

- Look for birds in their local area and describe their beaks.
- Match bird beaks to what they might eat.
- Create their own "beak designs" using art and imagination!

………….. And always remember ………………

It's all about the beak, about the bill, about the beak!

ABOUT THE AUTHOR

Meet Elysia Rainwood, an artist and writer passionate about conservation and environmental protection, particularly the preservation of the diverse animals that inhabit our wonderful Australian continent.

Growing up, Elysia always had an eye for detail, being very aware of such things as seasonal changes in the various environments she saw, from the depths of the lush, green rainforests to the sun-kissed beaches of the outer Great Barrier Reef islands. Her adventures fueled a desire to share the stories of the vulnerable and endangered.

Elysia writes non-fiction books and stories that resonate with the spirit of the tiniest of creatures and soar with the majesty of the wedge-tailed eagle. She likes to focus on the less well-known species to ensure that they and their habitats are not overlooked.

Elysia Rainwood's mission is clear:

To enchant young minds with the wonders of Australia's wildlife, fostering a generation of conservationists. Through interesting facts and engaging, reasonably accurate illustrations, she hopes to ignite a spark in the hearts of 4- to 10-year-old readers, inspiring them to protect the precious creatures that share our planet.

Along the way, Elysia also hopes to engage parents, caregivers, and educators who can pass this information along.

Made in the USA
Monee, IL
05 November 2026

67381724R00024